GLASGOW AND AYRSHIRE

A LANDSCAPE FASHIONED BY GEOLOGY

ISBN 1 85397 451 X

A CIP record is held at the British Library
W3K0406

Acknowledgements
Authors: Colin MacFadyen and John Gordon
Text on page 36: Alison Grant
Series editor: Alan McKirdy (SNH)

Photography:
BGS 16, 24; **Laurie Campbell** 19 bottom; **Lorne Gill/SNH** front cover, back cover, frontispiece, 6, 8 bottom, 9, 10, 15, 17, 21, 28, 29 left, 29 right, 30, 32, 33, 34 left, 34 right, 35 left, 35 right, 36, 37; **Patricia & Angus Macdonald/SNH** 20, 22, 23, 24, 26, 27, 31; **Colin MacFadyen** 4 right, 7, 38, 39; **Hunterian Museum and Art Gallery, University of Glasgow** 4 left, 12, 13 all; **Glyn Satterley** 25; **Eysteinn Tryggvason** 8 top.

Illustrations: Craig Ellery 2, 3, 5, 18; **Clare Hewitt** 14, 19 top; **Iain McIntosh** contents.

Front cover image:
The view north from Queen's Park on the south side of the city across central Glasgow, toward the Kilpatrick Hills, Dumgoyne and the Campsie Fells. Representing the eroded remnants of a Carboniferous lava plateau, these volcanic hills form a backdrop to the city which is built upon ice-moulded glacial deposits overlying Carboniferous sedimentary rocks. Locally derived sandstone has been used in the construction of the churches and other older buildings, such as Camphill Queen's Park Church, at the right of the picture.

Back cover image:
Fossil tree stump, Fossil Grove, Victoria Park, Glasgow

Further copies of this book and other publications can be obtained from:
The Publications Section,
Scottish Natural Heritage,
Battleby, Redgorton, Perth PH1 3EW
Tel: 01738 458530 Fax: 01738 458613
E-mail: pubs@snh.gov.uk
www.snh.org.uk

GLASGOW AND AYRSHIRE

A Landscape Fashioned by Geology

by

Colin MacFadyen and John Gordon

Glasgow is built on an ice-moulded landscape

Contents

Throughout the Glasgow and Ayrshire area, the effects of human activity are clearly evident in the landscape: rivers are bridged, the land is cultivated, and, in places, the rocks have been quarried and mined. This landscape, which is several hundred millions of years in the making, charts Scotland's journey across the face of planet Earth. This landscape fashioned by geology is not simply a backdrop to human history and development - it has been the fundamental factor in determining where and how people have lived and worked.

Glasgow and Ayrshire Through Time

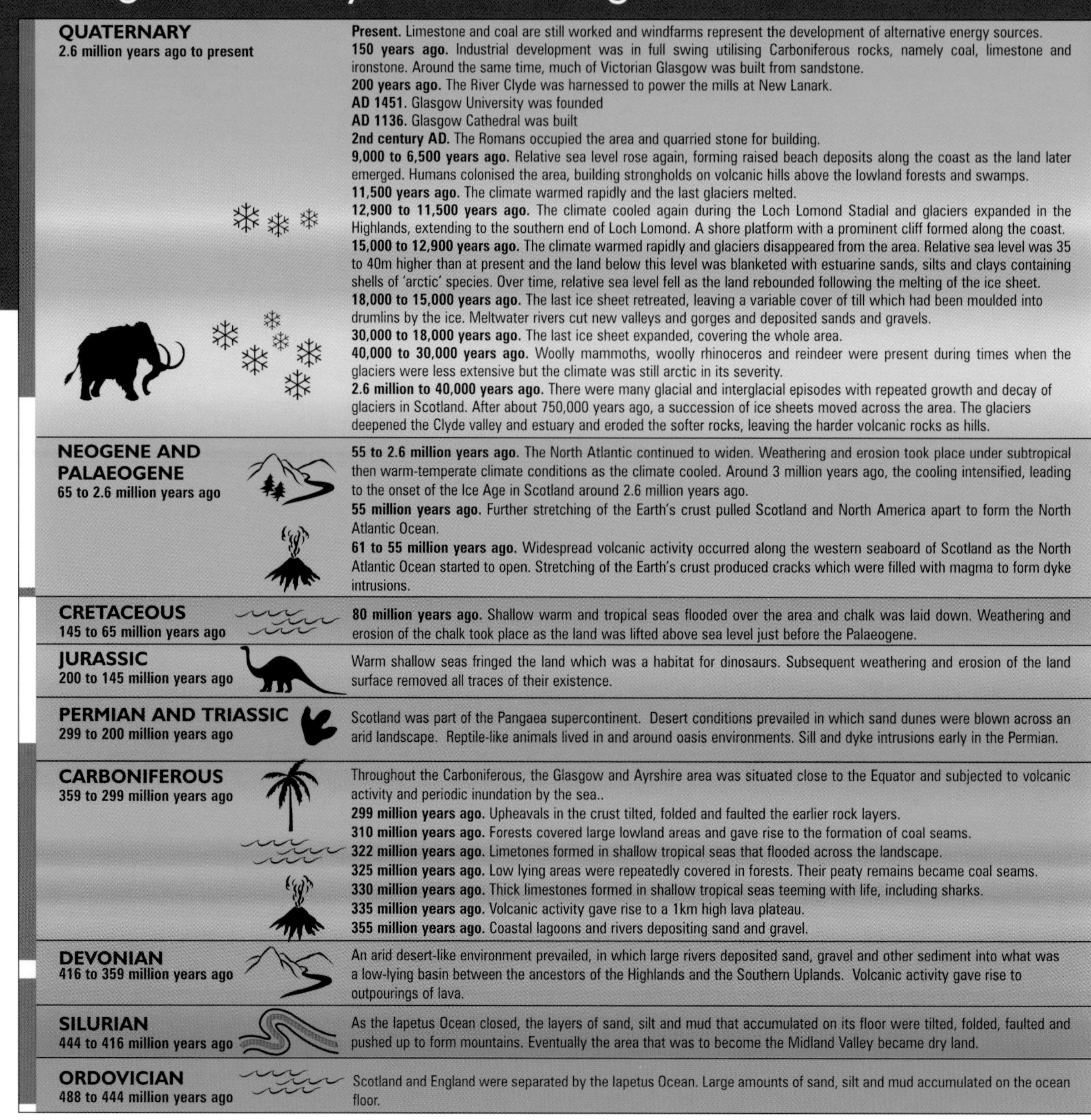

Period	Events
QUATERNARY 2.6 million years ago to present	**Present.** Limestone and coal are still worked and windfarms represent the development of alternative energy sources. **150 years ago.** Industrial development was in full swing utilising Carboniferous rocks, namely coal, limestone and ironstone. Around the same time, much of Victorian Glasgow was built from sandstone. **200 years ago.** The River Clyde was harnessed to power the mills at New Lanark. **AD 1451.** Glasgow University was founded **AD 1136.** Glasgow Cathedral was built **2nd century AD.** The Romans occupied the area and quarried stone for building. **9,000 to 6,500 years ago.** Relative sea level rose again, forming raised beach deposits along the coast as the land later emerged. Humans colonised the area, building strongholds on volcanic hills above the lowland forests and swamps. **11,500 years ago.** The climate warmed rapidly and the last glaciers melted. **12,900 to 11,500 years ago.** The climate cooled again during the Loch Lomond Stadial and glaciers expanded in the Highlands, extending to the southern end of Loch Lomond. A shore platform with a prominent cliff formed along the coast. **15,000 to 12,900 years ago.** The climate warmed rapidly and glaciers disappeared from the area. Relative sea level was 35 to 40m higher than at present and the land below this level was blanketed with estuarine sands, silts and clays containing shells of 'arctic' species. Over time, relative sea level fell as the land rebounded following the melting of the ice sheet. **18,000 to 15,000 years ago.** The last ice sheet retreated, leaving a variable cover of till which had been moulded into drumlins by the ice. Meltwater rivers cut new valleys and gorges and deposited sands and gravels. **30,000 to 18,000 years ago.** The last ice sheet expanded, covering the whole area. **40,000 to 30,000 years ago.** Woolly mammoths, woolly rhinoceros and reindeer were present during times when the glaciers were less extensive but the climate was still arctic in its severity. **2.6 million to 40,000 years ago.** There were many glacial and interglacial episodes with repeated growth and decay of glaciers in Scotland. After about 750,000 years ago, a succession of ice sheets moved across the area. The glaciers deepened the Clyde valley and estuary and eroded the softer rocks, leaving the harder volcanic rocks as hills.
NEOGENE AND PALAEOGENE 65 to 2.6 million years ago	**55 to 2.6 million years ago.** The North Atlantic continued to widen. Weathering and erosion took place under subtropical then warm-temperate climate conditions as the climate cooled. Around 3 million years ago, the cooling intensified, leading to the onset of the Ice Age in Scotland around 2.6 million years ago. **55 million years ago.** Further stretching of the Earth's crust pulled Scotland and North America apart to form the North Atlantic Ocean. **61 to 55 million years ago.** Widespread volcanic activity occurred along the western seaboard of Scotland as the North Atlantic Ocean started to open. Stretching of the Earth's crust produced cracks which were filled with magma to form dyke intrusions.
CRETACEOUS 145 to 65 million years ago	**80 million years ago.** Shallow warm and tropical seas flooded over the area and chalk was laid down. Weathering and erosion of the chalk took place as the land was lifted above sea level just before the Palaeogene.
JURASSIC 200 to 145 million years ago	Warm shallow seas fringed the land which was a habitat for dinosaurs. Subsequent weathering and erosion of the land surface removed all traces of their existence.
PERMIAN AND TRIASSIC 299 to 200 million years ago	Scotland was part of the Pangaea supercontinent. Desert conditions prevailed in which sand dunes were blown across an arid landscape. Reptile-like animals lived in and around oasis environments. Sill and dyke intrusions early in the Permian.
CARBONIFEROUS 359 to 299 million years ago	Throughout the Carboniferous, the Glasgow and Ayrshire area was situated close to the Equator and subjected to volcanic activity and periodic inundation by the sea.. **299 million years ago.** Upheavals in the crust tilted, folded and faulted the earlier rock layers. **310 million years ago.** Forests covered large lowland areas and gave rise to the formation of coal seams. **322 million years ago.** Limetones formed in shallow tropical seas that flooded across the landscape. **325 million years ago.** Low lying areas were repeatedly covered in forests. Their peaty remains became coal seams. **330 million years ago.** Thick limestones formed in shallow tropical seas teeming with life, including sharks. **335 million years ago.** Volcanic activity gave rise to a 1km high lava plateau. **355 million years ago.** Coastal lagoons and rivers depositing sand and gravel.
DEVONIAN 416 to 359 million years ago	An arid desert-like environment prevailed, in which large rivers deposited sand, gravel and other sediment into what was a low-lying basin between the ancestors of the Highlands and the Southern Uplands. Volcanic activity gave rise to outpourings of lava.
SILURIAN 444 to 416 million years ago	As the Iapetus Ocean closed, the layers of sand, silt and mud that accumulated on its floor were tilted, folded, faulted and pushed up to form mountains. Eventually the area that was to become the Midland Valley became dry land.
ORDOVICIAN 488 to 444 million years ago	Scotland and England were separated by the Iapetus Ocean. Large amounts of sand, silt and mud accumulated on the ocean floor.

Brown bars indicate periods of time represented by the rocks and deposits seen in the Glasgow and Ayrshire area.

Geological Map of Glasgow and Ayrshire

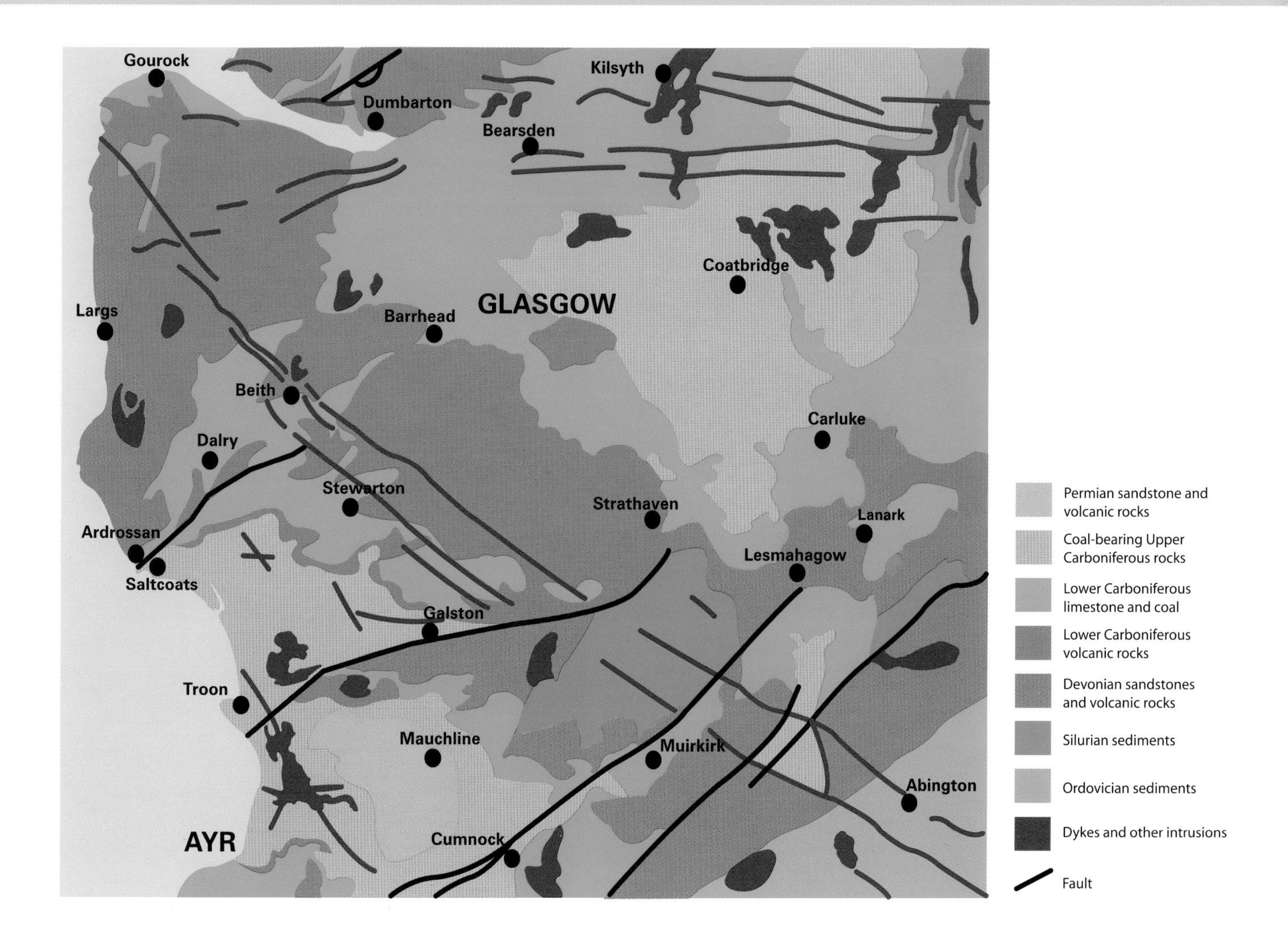

Scotland is United as an Ocean Closes

The story of the geology and landscape history of the Glasgow and Ayrshire area began around 450 million years ago when the oldest rocks were formed. This was during the Ordovician period of geological time, a time in Scotland's history when Scotland and England were separated by a wide ocean, called the Iapetus Ocean, which was the size of the modern-day Atlantic.

The area that was to become northern Scotland lay south of the Equator, on the coastal margin of a continent called Laurentia that also included Greenland and North America. As a consequence of plate tectonics, also known as 'continental drift', this continent collided with another continent that carried England.

The collision led to the disappearance of the Iapetus Ocean and brought together the various component parts of the Earth's crust that now form the foundations of the country. This union between northern Scotland and England occurred during the Silurian period and was both a destructive and constructive event leading to the formation of central and southern Scotland.

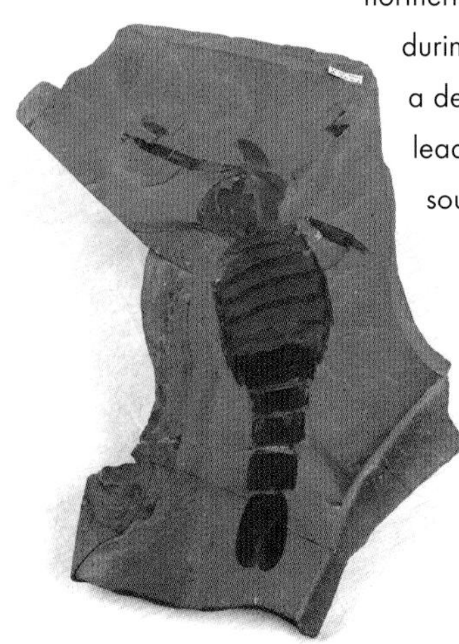

A Silurian eurypterid (fossilised scorpion-like sea creature)

Silurian rocks near Lesmahagow

Diagram showing the Midland Valley of Scotland late in the Silurian period after the closure of the Iapetus Ocean

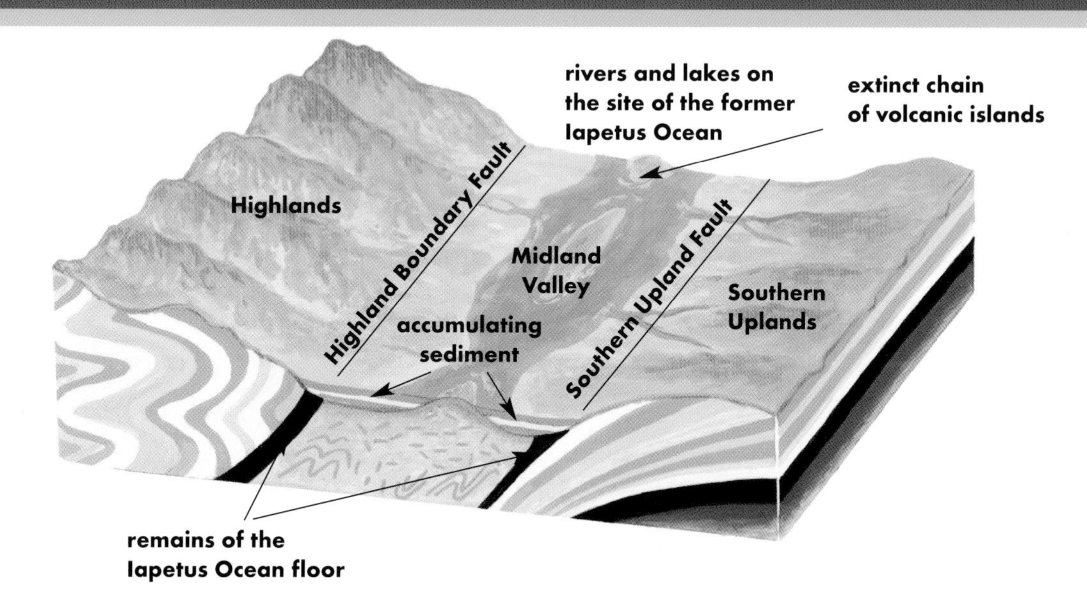

A chain of volcanic islands that lay between the continents containing northern Scotland and England, was caught up in the collision and became the foundation of the Midland Valley, or Central Belt, of Scotland. The ocean floor on either side of this volcanic island chain sank into the Earth's mantle, beneath the continental margins. The ocean floor sediments which had accumulated at the edge of Laurentia, north of the volcanic islands, were folded, deformed and baked to form the Dalradian metamorphic rocks of the Scottish Highlands. As the ocean south of the volcanic islands was swallowed up, the sand and mud on its floor was scraped off and piled up to form the Southern Uplands. Today the remnants of the ocean floor can be found along the margins of the Midland Valley.

Faults produced during the continental collision defined the margins of the Midland Valley: the Highland Boundary Fault to the north and the Southern Upland Fault to the south. The Midland Valley sank downwards along these faults and became a 'basin' into which rivers drained and lakes formed.

Silurian sedimentary rocks occur in the area between Lesmahagow and Muirkirk. They represent sand, silt and mud that accumulated in the closing Iapetus Ocean, then lagoons, and eventually lakes, as the area became dry land. In these watery environments some of united Scotland's first animals lived. They included primitive fish that lacked jaws and were akin to the modern eel-like lamprey. Other animals included large scorpion-like creatures known as eurypterids, some types of which grew up to two metres in length. The fossil remains of these early inhabitants of Scotland are found within the sedimentary rock layers.

A Desert Between Mountains

Red sandstone and conglomerate layers of Devonian age at Largs

Devonian conglomerate at Portencross, north-west of Ardrossan, laid down by a fast-flowing river in a desert-like environment

During the Devonian period, the crustal components forming Scotland became 'welded' together and the foundations of the country were complete. Scotland, however, continued to move with the drift of the continents and was heading north towards the Equator.

Although the environment would have been desert-like, vast braided river systems on the scale of the modern-day Ganges flowed into the Midland Valley basin, carrying boulders, gravel, sand and other sediment, which built up layer-upon-layer. In time, these loose sediments became consolidated to form the conglomerate and sandstone beds which are indicative of this period. Many of these rocks have a distinctive red-brown colouration, caused by a rusty coating of iron on the constituent sand grains. Sometimes known as the 'Old Red Sandstone', these Devonian rocks are well exposed along the Ayrshire coast and also in the Lanark area.

The ongoing drift of the continents resulted in upheavals in the newly formed foundations of Scotland during the middle part of the Devonian period. This disrupted the river systems and tilted newly formed rock layers. Late in the Devonian, sediment was once again deposited in the Midland Valley basin. This was derived from the erosive action of water and wind on the ancestors of the Highlands and Southern Uplands. The persistence of desert-like conditions at this time is illustrated by the occurrence of sand dunes in the rock record. The dunes would have blown across the surface of the basin as the whole area gradually sank down towards sea level at the end of the period.

Disturbance of the Earth's crust during the Devonian helped continue the formation of the low-lying Midland Valley and also gave rise to volcanic activity, particularly at the southern margin of the area at Galston. Volcanic activity was to be a key feature of the development of the landscape in this area for nearly 100 million years.

Volcanoes Shape the Landscape

At the beginning of the Carboniferous period around 355 million years ago, Scotland was drifting close to the Equator, and the desert-like environment gradually gave way to more tropical conditions. Rivers once again deposited sediment in the low-lying areas which, as result of subsidence of the crust beneath, were periodically flooded by the sea with the formation of lagoons. A battle for supremacy in the Glasgow and Ayrshire landscape between the dry land and the sea was to be the story for much of the Carboniferous period.

A major episode of volcanic activity in the Glasgow area, resulting from stretching of the crust, interrupted this battle for around 5 million years. During the volcanic activity, layer-upon-layer of lava erupted onto the landscape to form a lava plateau. Volcanic ash, which indicates more-explosive activity, was also produced.

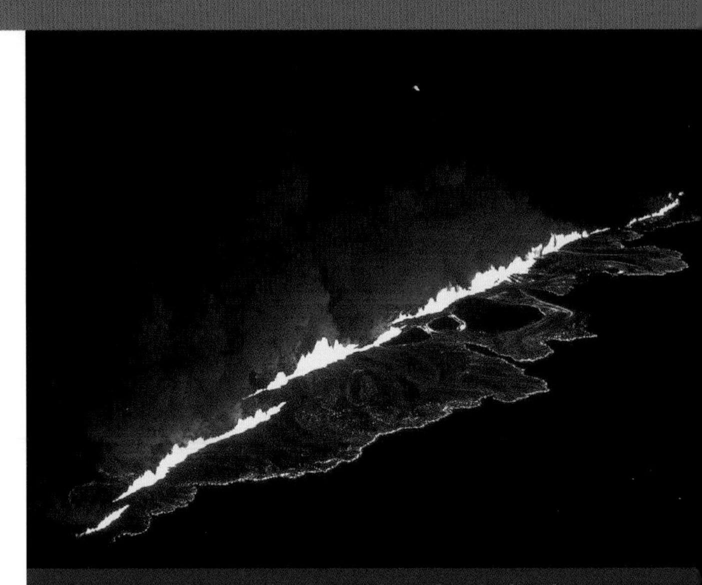

A fissure-style eruption in modern-day Iceland

A red-coloured 'fossil' soil between two lava flows at Loanhead Quarry near Beith. The top lava flow is about ten metres thick

Layer-upon-layer of lava underlies the landscape at Duntocher near the Erskine Bridge. The flows are tilted toward the right

The lava plateau reached a thickness of over a kilometre, the flows originating from eruptions along cracks or fissures in the crust in the Dumbarton to Kilpatrick Braes area. Although there was a thick accumulation of lava in the Glasgow area, only a few of the flows reached as far south as Ardrossan and beyond.

Known as the Clyde Plateau lavas, and composed mostly of basalt, these volcanic rocks form much of the high ground of the Glasgow and Ayrshire area and can be seen where rock is exposed between Greenock and Strathaven. The Renfrewshire Hills, the Gleniffer Braes, the Beith-Barrhead Hills and the Cathkin Braes are important landscape features, all fashioned from the early Carboniferous basalt lavas. Away from these areas of high ground, the lavas underlie younger rock layers.

Lava flows were also erupted from volcanic vents such as Dumbarton Rock. Such vents have been found to contain fragments of rock that came from the lower part of the Earth's crust. These provide a valuable insight into the rocks deep down in the crust below the Midland Valley.

The volcanic activity was not continuous and hundreds or even thousands of years may have elapsed between eruptions. This resulted in the weathering of the tops of the lava flows, and the formation of soils under the hot, moist conditions. Loanhead Quarry near Beith shows a good example of a layer of soil between two lava flows. Elsewhere, thick layers of volcanic ash can be found between some flows. This is illustrated at Boylestone Quarry near Barrhead, where three lavas and a thick bed of ash occur.

Following the end of the volcanic activity, weathering and erosion of the lava plateau took place. This, together with the variable thickness of the lava flows, produced an uneven landscape that affected the distribution of land and sea during the remainder of the early Carboniferous.

Sun, Sand and Sea

Layers of limestone in Trearne Quarry near Beith, formed in a shallow tropical sea

Following the eruption and then partial erosion of the Clyde Plateau lavas, the environmental conditions that existed before the volcanic episode were re-established in the landscape. Initially sediment such as mud and sand, derived from the weathering of the lavas, covered the uneven lava plateau. Eventually the battle for supremacy resumed between the land and sea with the encroachment of a shallow tropical sea onto the plateau. By 330 million years ago, similar environmental conditions prevailed over the whole area, as most of the lava plateau became submerged.

The shallow sea, which was akin to the modern-day Bahamas, repeatedly flooded the landscape over the next 10 million years. Limestone formed through chemical precipitation of calcium carbonate in the warm seawater and from the accumulation of the shells and skeletons of sea creatures. Lagoons formed in the coastal areas. At other times, rivers carrying sand, silt and mud into the sea, led to the formation of deltas and new land surfaces. Dense equatorial forests thrived and died on the swampy land surface on top of the deltas, and gave rise to peat, which, in time, became coal. Later, the sea would once again flood the land, repeating the cycle of environmental change.

Over millions of years, this constant switching from open shallow sea and coastal lagoons to swamp land and river conditions, then back again to marine conditions, gave rise to repeating rock layer cycles of limestone, mudstone, siltstone, sandstone, seatearth and coal. Under certain conditions, ironstone was formed, usually at the bottom of shallow-water lagoons.

Changing sea levels during Carboniferous times resulted in cycles of environmental change, with the formation of different rock types. One such cycle is shown here

Environment	Rock type	Fossils
tropical forest	coal/peat	plant stems
landsurface (soil)	fireclay/seatearth	plant roots
top of a delta with river channels	sandstone with cross-bedding	drifted plant remains
shallow water, muddy front of a delta	siltstone and sandstone	burrows made by sea floor animals and marine or freshwater mussel shells
falling sea level with approaching delta; offshore muddy water	mudstone with ironstone/ limestone nodules	marine shells including mussels
offshore deeper, clear water	limestone	shells, sea-lillies (crinoids), corals, shark teeth
offshore muddy water	mudstone	mussel shells
lake/estuary	siltstone and sandstone	freshwater or estuarine mussels, burrows
tropical forest	coal/peat	plant stems
landsurface (soil)	fireclay/seatearth	plant roots

Shark-infested Lagoons

Reconstruction of Bearsden during early Carboniferous times

A fossil shark ***Akmonistion zangerli*** **found at Bearsden as reconstructed left**

Dibunophyllum **- solitary coral**

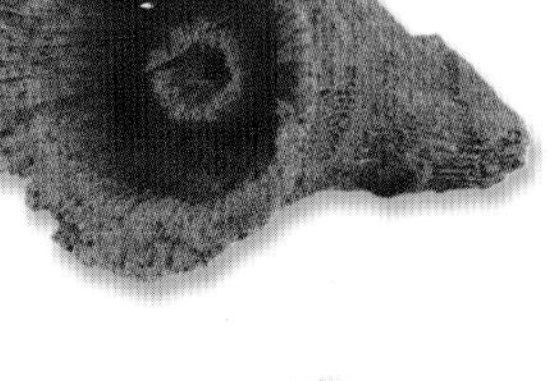

Lingula **- two shelled marine animal**

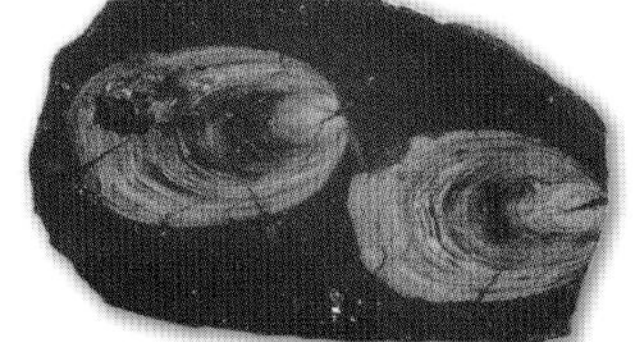

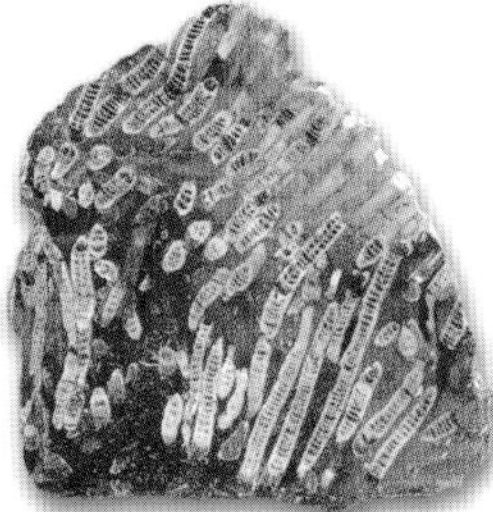

Lithostrotion **- colonial coral**

The fossil remains of creatures that thrived in the Carboniferous seas of Scotland are particularly well preserved in the limestones of Ayrshire. At Trearne Quarry near Beith, there are the fossil remains of a rich and diverse marine fauna, including corals, sponges, the ancestors of molluscs such as squid, oysters, sea snails and even jellyfish. The occurrence of sharks' teeth indicates that there were also large marine predators in the ancient tropical sea.

Close to the land, a lagoon environment that developed around 330 million years ago in the area that is now Bearsden was home to a variety of sea creatures, including at least fourteen different species of fish, some of which were sharks.

Bearsden is one of the best and most important fossil shark sites in the world and many significant discoveries have been made there in recent years. Fossil collector Stan Wood, working with the Hunterian Museum in Glasgow, first discovered the site whilst walking his dog in the early 1980s. In addition to complete and superb specimens of sharks, the fossil remains of several unique ray-finned fish and shrimp-like crustaceans have also been found. The preservation of these fossils is so good that even the muscles and blood vessels are displayed. Some of the sharks, like the one illustrated here, had a peculiar triangular shaped, tooth-covered structure termed a 'brush-organ' attached to the back of the head. Probably a modified dorsal fin, the brush was likely to have been used in the mating display of these sharks. One species of shark, akin to modern-day 'ratfish', had crushing teeth that enabled it to eat molluscs and other shellfish from the floor of the lagoon. All was not peaceful in this tropical environment, however, as volcanic activity rumbled on in North Ayrshire at this time.

Ancient Swamps and a Landscape Under Strain

Coal swamp reconstruction

Key: a - dragonfly b - *Lepidodendron* (scale-tree) c - Tree fern d - *Calamites* (horse-tail) e - amphibian

In the second half of the Carboniferous period, the Glasgow and Ayrshire area was dominated at intervals by swamp and river environments. These were periodically drowned by the sea but not to the extent that significant limestones developed. This was the time when dense tropical forest covered large areas of the Midland Valley, giving rise to the economically important Coal Measures. Hot tropical conditions persisted, as Scotland was still in the vicinity of the Equator but slowly moving northwards.

There were three distinct low-lying basins in the area, where coal accumulated. These subsequently formed the coalfields of Central Scotland, Ayrshire and the Douglas valley. The three areas were separated by higher ground underlain by the lavas that had erupted some 50 million years earlier and defined in part by faults in the crust.

An impression of a Carboniferous forest environment may be gained from a visit to Fossil Grove in Victoria Park, Glasgow. In 1887, quarrying revealed eleven tree trunks in growth position, complete with the tops of their root systems showing. The fossil remains of tree trunks and branches also occur among the trunks on what would have been the floor of the ancient forest. These stumps are giant club mosses. Although older than the rocks of the Coal Measures, the part of the forest preserved at Fossil Grove provides a valuable insight into the spacing of trees in the coal forest environment.

Towards the end of the Carboniferous period, Scotland drifted north of the Equator. The great tropical forests died out and Scotland once again experienced hot and dry climatic conditions, becoming an arid desert.

By 300 million years ago, Scotland was part of the huge supercontinent called 'Pangaea', which comprised all the world's continents. The Glasgow and Ayrshire area remained low-lying and bounded by the mountains of the Highlands and the Southern Uplands.

The coming together of Pangaea, with the closure of an ocean to the south in the area that is now central Europe, created shock waves in the Earth's crust. These forces did not have a major impact in Scotland, although the newly formed Carboniferous rock layers were tilted and folded and, in places, broken by faults.

Ongoing tensions in the crust below the area continued to give rise to volcanic activity. In late Carboniferous and into very early Permian times, magma derived from partial melting of rock deep in the Earth was squeezed into cracks and between rock layers below the surface of the landscape. The magma formed horizontal, inclined and vertically orientated sheets, known as 'sill' and 'dyke' intrusions. Excellent examples of late Carboniferous - early Permian sills and dykes occur at Troon, Ardrossan and inland at Lugar.

The fossil remains of tree stumps at Fossil Grove, Victoria Park, Glasgow

A New Desert Landscape

During the Permian period, Scotland drifted further northwards into latitudes that are today occupied by the Sahara and Arabia. Around 260 million years ago, sand blew across a desert landscape to form dunes, and, like some of the earlier Upper Devonian desert deposits, they had a red colouration due to a rusty coating of the sand grains. These younger desert sandstones have become known as the 'New Red Sandstone', to distinguish them from the deposits of the Devonian period.

The remains of the red sand dune deposits of the Permian period occur at Mauchline. Here, hundreds of metres of desert sandstone accumulated as dunes built up on a foundation of several hundred metres of Permian lava. The lavas formed from a localised centre of volcanic activity that developed on top of the Ayrshire coal forest. At Howford Bridge the red desert sand deposits are well exposed, together with lavas and volcanic ashes. The former Ballochmyle stone quarry at Mauchline showed evidence for the huge sand dunes that once existed across Ayrshire. Study of these ancient dunes has revealed that the wind blew across the Permian desert landscape predominantly from the east.

Ballochmyle Quarry as it appeared around 1921, providing a cross section through Permian sand dunes

Riverside exposure of the characteristically pink-coloured Mauchline sandstone at Howford Bridge, near Mauchline

During the Triassic period, the climatic conditions became more seasonal. There were lakes that periodically dried out, and by 210 million years ago the sea started to encroach onto the landscape. Around the lakes and in inter-tidal areas, the short-legged reptilian ancestors of the dinosaurs would have left their footprints in the soft sandy sediment.

During the Jurassic and Cretaceous periods, the Glasgow and Ayrshire area alternated between being either dry land or shallow sea. Dinosaurs would probably have lived in the area, inhabiting vegetated landscapes. Towards the end of the Cretaceous, a sea covered much of Scotland and a layer of chalk, representing the remains of huge numbers of marine plankton, formed on the sea bottom. The chalk would have been very similar to that which can be seen today on the south coast of England. However, at the end of the Cretaceous, the land rose above the sea and the blanketing layer of chalk, together with underlying layers of rock that would have preserved the remains of dinosaurs, was eroded away.

Almost from the time it was fully formed during the Permian period, the Pangaea supercontinent was splitting apart and new oceans, such as the Atlantic, started to form. By the Palaeogene, around 60 million years ago, the North Atlantic was beginning to open. This caused tension in the crust of western Scotland, giving rise to a line of volcanoes from Arran to St Kilda. Evidence of the stretching that split the crust is evident where magma rose up and solidified in the cracks to form dyke intrusions across the area. There are good examples of these dykes at the coast, for example at Largs and Saltcoats, where they cut through softer sedimentary rocks. Some now stand proud, like walls, because they erode less easily than the softer rocks around them.

Had the split in the crust occurred 200 kilometres to the east, the Glasgow and Ayrshire area would now be part of the landscape of North America!

The Onset of the Ice Age

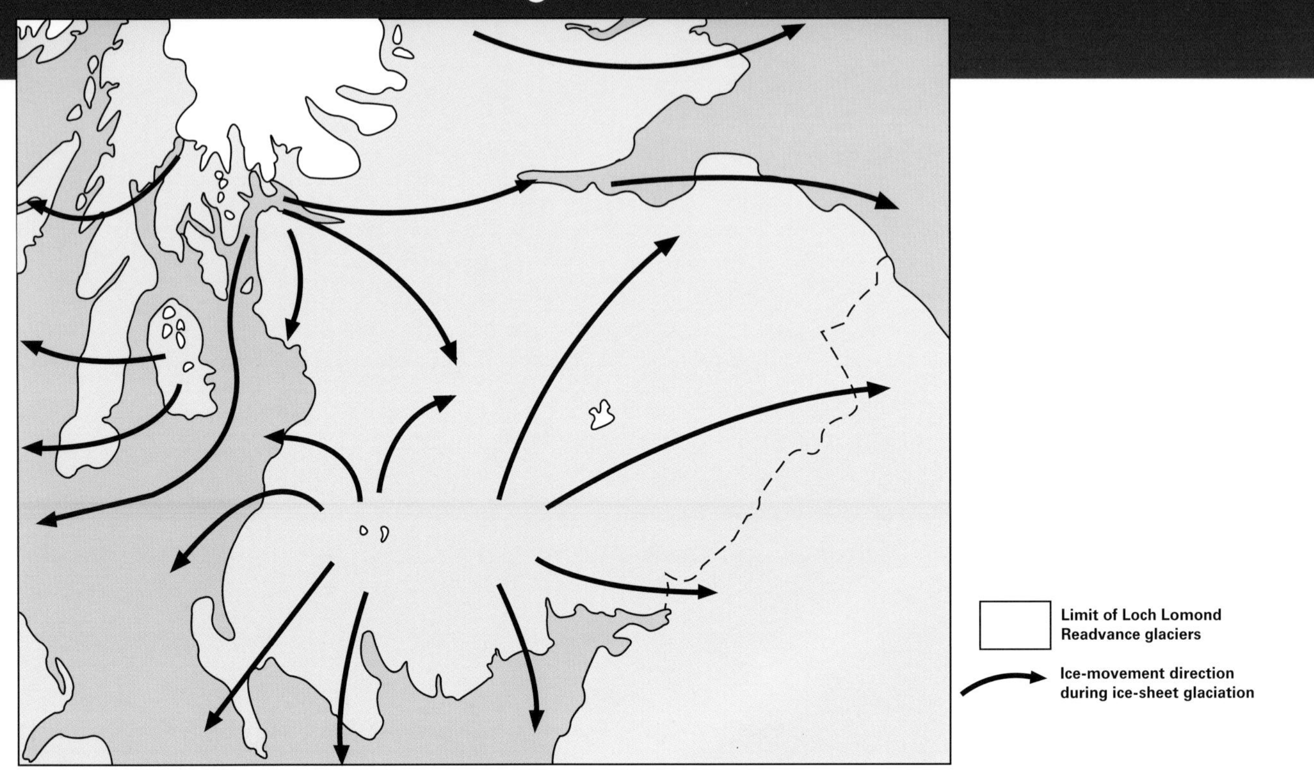

The diversity of landforms and landscapes in the Glasgow and Ayrshire area can be explained by the geological foundations and the weathering and erosion of the rocks over millions of years under a variety of subtropical and temperate climate conditions before the onset of the Ice Age around 2.6 million years ago, and since then by the action of frost and glaciers. The broad outlines of the landscape, as well as many of the details, owe much to the variety of hard, resistant rocks and softer, weaker rocks and the effects upon them of weathering and erosion. Although the general form of the landscape was largely in place before the onset of the Ice Age, the glacial period has left a particularly distinctive legacy in the landforms today.

During the Ice Age, the climate regularly swung back and forth between warm and cold conditions. In the colder episodes (glacials), glaciers existed in Scotland; during the warmer periods (interglacials), the climate was more like it is today and may even have been warmer. During the more intense glacials, ice sheets covered all of Scotland apart from the very highest summits. This probably occurred five or six times during the last 750,000 years, and during the many less cold episodes, smaller mountain glaciers existed in the Highland corries and glens. Today we are still in the Ice Age, albeit in a warmer, interglacial phase known as the 'Holocene'.

An artist's impression of the Ice Age environment

Each successive glaciation not only helped to shape the landscape, but also removed most of the deposits of earlier glaciations. Therefore much of the evidence we see today dates from the time of the last glacial period, between about 30,000 and 11,500 years ago. However, there are a few crucial sites where earlier deposits have fortuitously survived. Typically, these are buried layers of peat or soil, and the pollen, plant and insect remains that they contain provide a wealth of environmental information about vegetation and climate before the last glaciation. In parts of lowland Ayrshire and the lower Clyde valley, bones have been found of large mammals that are now extinct in Scotland, including mammoth, woolly rhinoceros and reindeer. These date from the time before the last major ice sheet glaciation and indicate a period of cold, non-glacial conditions like parts of the Arctic today.

Reindeer were present in Scotland during the Ice Age

Ice Shapes the Landscape

During successive glaciations, the glaciers deepened the lower parts of the Clyde and Kelvin valleys, forming rock basins that descend to over 70 metres below present sea level. These were later infilled with glacial and other deposits. Elsewhere, ice-streamlined landforms are common, such as the crag-and-tail features at Loudoun Hill in Ayrshire and the Necropolis in Glasgow. These formed where more resistent volcanic plugs or sills protected weaker sedimentary rocks on their lee sides from the full force of the glaciers.

During the last glaciation, glaciers expanded out from the south-west Highlands some time after about 30,000 years ago and reached their maximum extent around 22,000 years ago or slightly earlier. The ice extended south and eastwards from Cowal, Loch Fyne and Loch Lomond across the Midland Valley and southwards along the Firth of Clyde, which was then dry land because of the lower world sea level. The glaciers formed a vast sheet of moving ice which buried the local landscape to a depth of perhaps 1,400 metres. Glaciers also built up in the Southern Uplands and moved northwards across the southern and central part of the area, merging with the Highland ice and diverting the latter westwards across Ayrshire. This is reflected in the character of the glacial deposits in central and southern Ayrshire. Here, glacial deposits with erratics (far-travelled rocks) carried by the glaciers from the Highlands are overlain by deposits derived from the Southern Uplands.

The glaciers smothered the lowlands in a blanket of till – a chaotic mix of boulders, gravel and pebbles in a matrix of sand and clay picked up as the ice moved across the frozen wastes. The colour of the till varies according to the type of rocks that the glaciers flowed across and eroded; for example in the Glasgow area, the till is red where it contains Devonian rocks and grey where it contains Carboniferous rocks.

Drumlin landscape north of Glasgow

The ice also moulded the till to form distinctive areas of elongated low hills, or drumlins, in the Glasgow area and in lowland Ayrshire. The buildings of Glasgow University occupy prominent positions on the crests of the drumlins, while curving street patterns in certain areas, for example Maryhill and Mosspark, follow the layout of the landforms. The hilly drumlin topography of the city is also reflected in many of its place-names, such as Jordanhill, Maryhill, Hillhead and Drumchapel.

The drumlins are generally steeper on their western sides. Their orientations also show how the ice flow fanned out eastwards north of the Clyde and south-eastwards south of the Clyde as it moved out from the Highlands.

An unusual feature of the glacial till deposits in Ayrshire is the presence of sea shells, which indicates that the ice moved onshore from the Firth of Clyde at some stage during the glaciation.

The Clochodrick Stone is a large glacial erratic of basalt lava that was transported by a glacier to its present location near Lochwinnoch, north of Beith, from the Renfrew Heights to the north

Melting Ice Makes its Mark

Carstairs Kames are a series of esker ridges formed by glacial meltwater rivers. The former extent of the landforms has been significantly reduced by sand and gravel quarrying

About 15,000 years ago, the last major ice sheet melted as the climate warmed rapidly. The melting ice released vast quantities of meltwater which cut channels in the bedrock. These channels now form dry valleys, for example on the north side of Tinto Hill. On lower ground, the meltwater deposited large amounts of sand and gravel, notably in a belt from south-west of Lanark to north-east of Carstairs. These deposits often take the form of mounds (kames), ridges (eskers) and terraces, sometimes laid down in contact with the ice. Good examples of eskers occur at Eaglesham Moor, Kaims of Avon and Chapelton, but the most spectacular are the Carstairs Kames. These deposits form part of a major glacial drainage system directed north-east towards the Firth of Forth. Only after the lower Clyde became ice-free did the drainage resume a north-west route along the Clyde valley.

The Highland glaciers retreated towards the north-west, so that the Clyde estuary remained blocked by ice at a time when the ice in the Clyde valley had melted. As the glaciers retreated, lakes were dammed in the middle and lower Clyde valley ('Lake Clydesdale') and in the valleys of the Avon Water and the River Kelvin. Large thicknesses of sand and gravel were deposited in deltas in the lake around Loudoun Hill in the valley of the Avon Water, but these have now been quarried away.

The final stages of ice retreat were accompanied by the removal of the ice barrier across the Clyde estuary, allowing the ice-dammed lakes to drain, possibly in great floods. The sea also flooded into the Glasgow area up to an altitude of 35 – 40 metres above present sea level, forming a large embayment in the Paisley-Linwood area. This occurred because the land

Sand and gravel deposited in a glacial lake at Loudoun Hill has been extensively quarried

surface was still depressed from the weight of the ice. Thick beds of estuarine clays and silts were deposited containing cold water, or boreal, sea shells. These can be seen in stream sections at Geilston near Cardross. At this time, the sea also flooded the lower lying parts of the Ayrshire coast, forming a large embayment between Ayr and Ardrossan. Relative sea level then fell as the land rebounded.

The improvement in the climate, however, was short-lived and glaciers once more took hold in the Highlands some time after 13,000 years ago, during a brief cold period known as the Loch Lomond Stadial. These Loch Lomond Readvance glaciers produced many of the moraines that we see today in the Highland glens and the large end moraines around the southern end of Loch Lomond. The summits of the hills in the southernmost part of the Glasgow and Ayrshire area are extensively covered in frost-shattered debris and deposits formed by the slow downslope movement of the soil (solifluction), that probably date in part from this time. Solifluction also affected the steeper slopes of the drumlins in the Glasgow area. On Tinto Hill, where the vegetation has been eroded, superb stone stripes are actively forming today through freezing and thawing of the soil in winter.

Changes at the Coast

The Main Rock Platform at Seamill extends inland beneath postglacial raised beach deposits to a prominent cliffline

During the glacial periods, sea level fell as the expanding ice sheets locked up the world's freshwaters, and rose again during interglacials when the glaciers melted and released their water back into the sea. The level of the land also varied, sinking downwards under the weight of the growing ice sheets and rebounding when they melted. Such changes are now evident in raised beaches around Scotland's coastline, where they provide flat building land and the basis for some of our most famous links golf courses including Troon and Turnberry.

A shore platform and rock-cut backing cliff are conspicuous along much of the coastline of the Firth of Clyde and in the Clyde estuary. They are particularly well displayed between Largs and Inverkip, between West Kilbride and Ardrossan, at Heads of Ayr and around the Ardmore peninsula near Cardross. On Great Cumbrae the platform forms a raised bench running the whole way round the island. The platform is part of a feature known as the Main Rock Platform. It is tilted and dips southwards, becoming intertidal near Heads of Ayr, due to less crustal rebound in the south where the ice was thinner.

This platform and cliff were cut during the Loch Lomond Stadial through a combination of frost weathering, sea-ice and wave action, but may in part be older features that were re-occupied by the sea. The platform is often overlain by postglacial (Holocene) raised beach deposits, mainly of sand and gravel.

Later, about 9,000 years ago, the sea level rose again and large areas of the coastline were drowned. The sea flooded once more into the lower Clyde valley, forming a large embayment in the Paisley-Linwood area, submerging the ground below an altitude of about 7.5 metres. In the Irvine area in Ayrshire, the sea extended 4–5 kilometres inland from the present-day coast. Some of these changes can be seen in the deposits in the valley of the River Irvine. Here, a buried layer of peat indicates relatively lower sea levels before about 9,000 years ago. Rising sea levels, however, covered the peat with several metres of beach sands and gravels. Sea shells and even the bones of stranded whales have been found some distance inland in these raised beach deposits. The sea remained at a relatively high level until about 7,000 years ago, then fell to its present level.

Turnberry Golf Course is located on a raised beach

Rivers Reappear

The Clyde is the main river in the area. Rising in the Southern Uplands and flowing north-west through Glasgow to the Clyde estuary and the Firth of Clyde, the river has played a crucial role in the origin and development of the city.

The Clyde and other postglacial rivers (such as the Irvine and Ayr) inherited a landscape modified by glaciers. The last ice sheet had infilled the pre-existing river valleys with glacial deposits and, in some cases, buried them completely. When the ice melted, the rivers did not always follow their previous courses, but excavated new valleys, leaving the old channels buried. A good example is the Clyde near Lanark. Here, the river cut a new postglacial course through a narrow, seven kilometre long rock gorge, producing two spectacular waterfalls at Bonnington Linn and Cora Linn. Similarly, the Kelvin and the Mouse, tributaries of the Clyde, have carved deep new postglacial courses. From the late eighteenth century, the power of the Clyde at New Lanark was harnessed to drive cotton-spinning machines.

Elsewhere, the postglacial rivers re-worked the glacial deposits, meandering backwards and forwards across their floodplains until modern river engineering constrained their courses through canalisation and flood embankments. The River Clyde at its junction with the Medwin Water is a good example of an actively meandering river which has not been subject to major engineering works. Former positions of the river are revealed in the pattern of well-preserved abandoned channels, meander cutoffs and oxbow lakes on its floodplain. The layering of the sediments within the abandoned channels records the occurrence of past floods on the river.

The Clyde estuary between the Erskine Bridge and Dumbarton Rock

Meanders and abandoned channels of the River Clyde near its junction with the Medwin Water

The Clyde estuary begins down-river from the Erskine Bridge, where the narrow valley cut through resistant Carboniferous lavas widens out. Westwards, there is a sharp transition from the shallow estuary to deep marine waters near Greenock, where the estuary joins the glacially deepened drowned valleys emerging from the Highlands and which now form the Firth of Clyde. The estuary, like areas upstream, is infilled with thick deposits of till and estuarine sediments, exceeding 45 metres in depth, plugging the pre-glacial valley. The shallow water depth of the estuary is evident at low tide, when banks of mud and sand are exposed between Dumbarton and Greenock. Since the mid-eighteenth century, major efforts have been made to manage the estuary and to develop and maintain a deepwater channel for navigation into the heart of Glasgow. These engineering works helped sustain the industrial development of the city as ships increased in draught.

Exploitation of the Geological Resources

In terms of economic resources the Carboniferous was arguably the most important period in Scottish geological history. Three rock types formed during this time, namely coal, limestone and ironstone, were crucial for the industrial development of the area. The extraction of these geological resources goes back to a time before accurate records were kept, but it is thought that medieval ecclesiastical communities were involved in their utilisation. Although the level of activity has declined considerably, reserves of coal and limestone are still being exploited.

The coal seams formed during Carboniferous times literally fired the industrial revolution. Coal was used with another product of the Carboniferous environment, ironstone, in the iron and steel industry. Coal powered the steam-driven transport, the trains and ships, and heated homes. Coal was also used to burn limestone to produce lime for agricultural purposes and for the production of lime mortar and plaster used in the building industry.

Mine waste in an industrial landscape near Kirkintilloch

'Stoop-and-room' limestone workings at Baldernock, near Milngavie

Subsidence at Châtelherault

The fossil soils or seatearths found in association with the coal beds were also of economic importance. Tropical weathering in the Carboniferous environment washed the minerals out of the soil leaving a muddy residue, called kaolinite clay, rich in aluminium. Bricks of baked kaolinite were used to line furnaces. An example is the Ayrshire Bauxitic Clay found in the Dalry to Kilmarnock area, which is one of the highest quality fireclays in the UK, worked formerly at the Monkcastle Mine in Dalry. Mudstone also formed in the ancient Carboniferous environment, and, as a by-product of coal mining, was often used in brick making.

The extraction of these resources has left a legacy of mines, quarries and dumps that have affected the landscape. Some of the dumps form artificial hills including one near Kirkintilloch. These have largely been removed for use as fill, for example in road construction work, or smoothed out and landscaped.

At the Linn of Baldernock east of Milngavie, so-called 'stoop-and-room' workings in limestone can still be seen. This was a method of mining that involved removing some of the resource but leaving pillars to hold up the roof. Limekilns are scattered across the landscape and are testament, in the agricultural context, to the burning of local limestone to provide lime for improving soil quality.

The collapse of old mine workings below ground has led to subsidence at the surface in various places. Hamilton Palace, the home of the Dukes of Hamilton, was demolished in the 1920s following structural damage resulting from subsidence of the ground beneath it. Châtelherault, formerly part of the Hamilton's estate within the Châtelherault Country Park, is still standing but bears witness to ground subsidence.

Human Activity Continues to Fashion the Landscape

During the twentieth century, the deep mining of coal became uneconomic and today there are no underground coal mines operating in the area. However, the extraction of coal does continue through opencast working, where it is removed by stripping away the overlying rock layers. Opencast mining can allow the extraction of coal from areas where the geology is complex and virtually impossible to mine conventionally due to the presence of faults and waterlogged old workings. Working from the ground surface downwards also allows for the extraction of several layers, or seams, of coal rather than having to dig a separate mine tunnel into each seam, as in conventional coal mining.

At Glenbuck by Muirkirk, coal seams within the Douglas Coalfield are being worked by opencast coal methods. In such opencast operations, the topsoil is stripped off the bedrock and stored in mounds. The rock which overlies the coal is called the 'overburden'. It is removed to reveal the coal and is used directly to 'backfill' the areas of the opencast mine that have been 'coaled-out'. Once the accessible and economic coal reserve has been removed and the site has been backfilled, the soil is replaced and efforts are made to have the area restored to marry in with the surrounding landscape.

Other large modern quarries exist for the extraction of Carboniferous-age igneous rocks, which tend to be strong and durable. Usually the rock is crushed into small fragments known as 'aggregate' for use in the construction industry. Aggregate continues to be worked in the Swinlees Quarry on the outskirts of Dalry. The rock is an unusual type of lava known as 'rhyolite', that fills a volcanic vent, which was erupting at the same time as the Clyde Plateau lavas.

Sand and gravel deposited by glacial meltwater rivers have been exploited along the Clyde and Kelvin valleys. At Ardeer in Ayrshire, blown sand in dunes formed after the Ice Age has been extracted, the sand being used for glass-making and foundry work.

Extraction of rock for use as aggregate at Barr's Swinlees Quarry near Dalry

Opencast coal workings in the Muirkirk area

Buildings from Sand

View over the lawn at Dean Castle, Kilmarnock

The built environment is another human aspect to the landscape, and for millennia people have used stone in the construction of dwelling houses and other buildings. Wherever possible the stone was sourced locally, literally within metres of the building site. The quarried building stone used throughout the Glasgow and Ayrshire area was almost invariably sandstone derived from the Devonian, Permian and, more commonly, the Carboniferous rock layer sequences. The colour of the building stone is usually a good indicator of its age and therefore how it was formed. Most of the well-known red sandstones of Ayrshire were formed in the arid and desert-like environments of either the Devonian or the Permian periods. White, cream or buff-coloured sandstones from the Carboniferous period are mostly river channel deposits.

At Bearsden, in the second century, the Romans utilised the local Carboniferous sandstone in the construction of a bathhouse. Large, flat slabs of rock were well suited for paving the floors of the structure, whilst smaller blocks were used in the construction of the walls. This early construction work would have required a degree of quarrying to provide such large slabs.

Many of the area's other historic buildings were constructed from sandstone obtained within metres of their locations. Dean Castle in Kilmarnock, which dates from the fourteenth century, and was for centuries one of Scotland's most impregnable baronial strongholds, and Portencross Castle illustrate the use of locally sourced building stone.

In the nineteenth and earlier part of the twentieth centuries, sandstone from local quarries was used in the construction of prestigious buildings and housing, including tenements. The Bishopbriggs and Giffnock Sandstones were the two most important building stones in the Glasgow area; both were extensively quarried and then mined to build most of Victorian Glasgow.

The construction of urban vernacular houses and other buildings in the past would, for the most part, have utilised rubble derived from the nearest rivers and from fields. However, when resources permitted, sandstone would have been quarried and transported for use as dressed stone for lintels, door jambs and quoin stones on corners.

Flagstones of local Carboniferous sandstone form the floor of the Roman bathhouse at Bearsden

The Built Environment Tells a Tale

The use of 'imported' stone in the construction of some of the prestigious buildings of Glasgow and the surrounding area documents the increasing wealth during the Industrial Revolution and subsequently. The increased 'geodiversity' of the city centre also reflects the development of communications, principally the railways, powered by coal. In Glasgow's St Vincent Place, the building at number 24 was built using red-coloured Permian sandstone from Mauchline. This sandstone was quarried from Ballochmyle Quarry pictured on page 16. Nearby at 86-90 St Vincent Street, there is a white or off-white building stone that was transported to Glasgow from Dorset on the south coast of England. This is Portland stone, a limestone formed in shallow tropical seas during the Jurassic period. Another 'alien' rock type used in the building of Glasgow can be seen at 34-38 West George Street. Here, pink-coloured Peterhead granite has been used in the construction of the pillars and balustrades. Unlike the Mauchline sandstone and the Portland stone, which are sedimentary rocks, the Peterhead granite is a coarse grained igneous rock derived through the slow cooling of molten magma at depth in the crust.

Study of these building stones and others in the area can reveal the story of their formation. The desert origin of the Mauchline Sandstone is apparent from the grains of sand from which it is made. The grains are rounded and have a rusty iron coating, indicative of being blown across a dry desert floor under oxidizing conditions. The layering evident in this Permian sandstone represents the original layering within mighty sand dunes.

Sandstone at Châtelherault shows evidence of being formed in a river

Ripple marked sandstone, produced in shallow running water, forms the floor of the Bearsden bathhouse

The lower storey pillars at 34-38 West George Street, Glasgow, are made of Peterhead granite

24 St Vincent Place, Glasgow, was built using red Permian sandstone from Mauchline

In contrast, the light-coloured Portland stone contains fragmented fossil remains of the skeletons and shells of marine creatures which tell the tale of its formation in a shallow tropical sea. The interlocking mineral grains of the Peterhead granite are testimony to its molten origin. All the major granite-forming minerals can be seen with the naked eye, principally clear quartz, pink feldspar and dark mica.

Evidence for the origins of stones used in the construction of other buildings highlighted in this book can, with careful observation, be revealed. Blocks of the locally sourced red sandstone from which Châtelherault is built, show layers known as 'cross-bedding', produced under flowing water in a Carboniferous river environment. The sandstone slabs utilised in the second century Roman bathhouse at Bearsden have preserved ripple marks of the type found today in any sandy shallow marine or river environment. Given the similarity between these ancient ripples and their modern day equivalents, it is worth speculating for a moment whether the Roman stonemasons who constructed the bathhouse were able to read the tale of the ancient structures in the rock.

The Landscape Today *by Alison Grant*

The Finnieston Crane on Stobcross Quay in Glasgow's docks, a 59 metre high cantilever crane built in 1932 to load heavy machinery onto freighters and transfer boilers and engines into new vessels

"The Clyde made Glasgow and Glasgow made the Clyde" This local saying embodies the relationship between the geography, culture and economics of the vast city of Glasgow and the river which flows through its heart.

During the eighteenth and nineteenth centuries, the city and its extensive hinterland were the focus of a vibrant and rapidly changing society. The deep waters of the Clyde were the catalyst for manufacturing enterprise and international trade, while the city expanded to accommodate a rising population, with streets and parks often arranged to make the most of extensive views from the drumlins. The coastal towns became holiday resorts, and the river was used for pleasure boats as well as commercial traffic.

Underpinned by an impressive industrial infrastructure, the rich natural resources were exploited across the surrounding landscape. The increased wealth generated by this industrial endeavour was used to develop grand mansions and an improved agricultural landscape with hedges and field trees. While many of these estates are either built over or in decline, some of the parks and policies now form the core of today's popular country parks.

The landscape today is still in the grip of change. As quickly as the legacy of past industrial expansion disappears, new land uses such as forestry, opencast mining and wind power take their place.

Contrasting architectural styles on the seafront at Ardrossan, with Ardrossan Windfarm on the hills behind. The earlier houses have been built using sandstone derived locally

Sites of Special Scientific Interest (SSSIs)

Lynn Spout SSSI near Dalry

There are around 34 Sites of Special Scientific Interest, or SSSIs, in the Glasgow to Ayr and Clyde valley area, reflecting the national and international importance of the geological and landform heritage of the area. Such sites include Fossil Grove, which provides an invaluable and precious insight into the environment and growth of trees in a Carboniferous forest around 325 million years ago. Fossil Grove was one of the first locations anywhere in the world to be afforded protection and conserved for future generations.

Other sites include Trearne Quarry at Gateside near Beith. The huge variety of well-preserved fossils contained within the limestone beds at Trearne have provided an outstanding insight into the complex ecosystem within the shallow tropical sea that existed over the area around 330 million years ago. The rock sequence at Rouken Glen SSSI on the outskirts of Glasgow, is of national importance as it provides a valuable insight into the geology of the Central Scotland Coalfield.

The Clochodrick Stone SSSI, which is one of the smallest protected sites in Scotland, is important for illustrating the movement of ice across the area during the Ice Age. The Carstairs Kames are an outstanding example of an esker system. There are also river SSSIs, nationally important for river landforms and ongoing dynamic processes, such as the Falls of Clyde and the Clyde Meanders.

Geological and landform sites are often just as vulnerable to changes in land use as wildlife habitats. For example, sand and gravel extraction around Carstairs has completely destroyed parts of the irreplaceable Carstairs Kames esker system and the adjacent landforms. Fossil sites in the vicinity of Lesmahagow have been targeted by ruthless and irresponsible collectors, which has threatened to destroy an internationally significant fossil resource. Appropriate management will ensure that these important sites are safeguarded for the benefit of future generations.

RIGS - Regionally Important Geological and Geomorphological Sites

Scottish Natural Heritage is concerned not only with the conservation, enhancement and interpretation of the statutory network of Sites of Special Scientific Interest (SSSIs), but also with the promotion of geoconservation in the wider countryside and at a local level. This is delivered through support for the enhancement and interpretation of Regionally Important Geological and Geomorphological Sites (RIGS). RIGS are a non-statutory designation and are defined according to four nationally agreed criteria: for their educational, research, historical and aesthetic value. To date in Scotland, the emphasis has very much been on the designation of sites for use in education and interpretation.

Currently there are six RIGS groups in Scotland: Fife, Highlands, Lothian and Borders, Stirling and Clackmannan, Tayside, and Strathclyde. Representatives on these groups include: local authority planners, rangers, museum and education services; Scottish Natural Heritage (Advisory and Area staff); British Geological Survey; Scottish Wildlife Trust; Scottish Rural Property & Business Association; universities; and amateur volunteers.

How does a local RIGS group work? The answer is: in many different ways. Generally they have a mainly professional committee that meets quarterly and a volunteer group that meets monthly. The committee monitors progress, designates RIGS, and directs the efforts of the volunteer group in producing interpretation leaflets and posters and visiting existing or potential RIGS sites.

More information on Scottish RIGS may be found at www.scottishgeology.com

Rock exposure on the beach at Saltcoats, possibly a candidate for interpretation by the local RIGS group

Scottish Natural Heritage and the British Geological Survey

Scottish Natural Heritage is a government body. Its aim is to help people enjoy Scotland's natural heritage responsibly, understand it more fully and use it wisely so that it can be sustained for future generations.

Scottish Natural Heritage
Great Glen House, Leachkin Road
Inverness IV3 8NW

SCOTTISH NATURAL HERITAGE

The British Geological Survey maintains up-to-date knowledge of the geology of the UK and its continental shelf. It carries out surveys and geological research.
The Scottish Office of BGS is sited in Edinburgh. The office runs an advisory and information service, a geological library and a well-stocked geological bookshop.

British Geological Survey
Murchison House
West Mains Road
Edinburgh EH9 3LA

British Geological Survey
NATURAL ENVIRONMENT RESEARCH COUNCIL

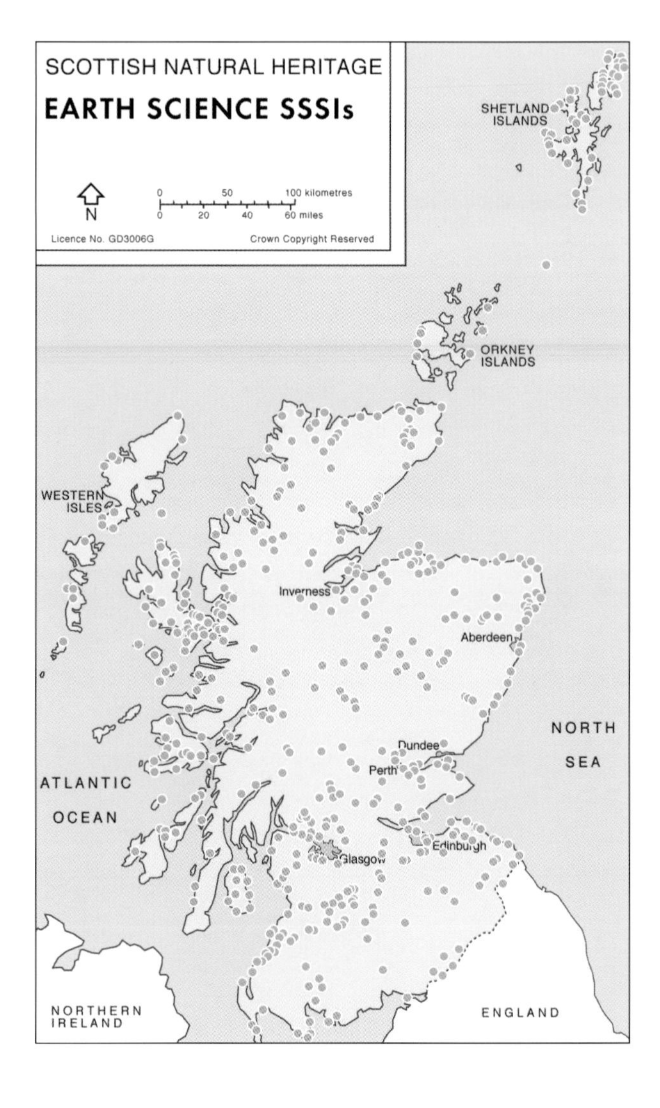

Remember the
Geological Code!
SCOTTISH
NATURAL
HERITAGE
GEOLOGISTS' ASSOCIATION
1858
No need to hammer indiscriminately!
Never collect from walls or buildings.
Keep collecting
to a minimum:
remove fossils, rocks
or minerals only
when essential for
serious study.
And remember to
refer good finds
to local
museums.
The leader of a field
party should ensure
that the spirit of the
code is upheld.
Always seek
permission before
entering
private
land.
No one has the right to "dig out" any site.
Try to leave the site as you found it!
Don't litter fields or roads with rock
fragments, and avoid disturbing plants or wildlife.
Back fill excavations where necessary to
avoid injury to people or animals.
Be considerate, and do not make things
more difficult or hazardous for others
coming after you.
Don't disfigure rock surfaces with brightly
painted numbers, symbols or clusters
of core-holes.
SAFETY FIRST!
✔ Wear protective goggles when hammering.
✔ Wear safety hats in quarries or below cliffs.
✔ Avoid loosening rocks on steep slopes.
✗ Do not get cut off by the tide.
✗ Do not enter old mine workings or cave systems.
✗ Do not interfere with machinery in quarries.
● Remember, you are one of several hundred geologists visiting this area every year — so your behaviour does matter.
● Please observe the code, so that others can also enjoy the great scenery, geology, and ecology here!
Published by Scottish Natural Heritage, 1996

Also in the Landscape Fashioned by Geology series...

Arran and the Clyde Islands
David McAdam & Steve Robertson
ISBN 1 85397 287 8 pbk 24pp £3.00

Cairngorms
John Gordon, Rachel Wignall, Ness Brazier, and Patricia Bruneau
ISBN 1 85397 455 2 pbk 52pp £4.95

East Lothian and the Borders
David McAdam & Phil Stone
ISBN 1 85397 242 8 pbk 26pp £3.00

Edinburgh and West Lothian
David McAdam
ISBN 1 85397 327 0 pbk 44pp £4.95

Fife and Tayside
Mike Browne, Alan McKirdy & David McAdam
ISBN 1 85397 110 3 pbk 36pp £3.95

Glen Roy
Douglas Peacock, John Gordon & Frank May
ISBN 1 85397 360 2 pbk 36pp £4.95

Loch Lomond to Stirling
Mike Browne & John Mendum
ISBN 1 85397 119 7 pbk 26pp £2.00

Mull and Iona
David Stephenson
ISBN 1 85397 423 4 pbk 44pp £4.95

Northwest Highlands
John Mendum, Jon Merritt & Alan McKirdy
ISBN 1 85397 139 1 pbk 52pp £6.95

Orkney and Shetland
Clive Auton, Terry Fletcher & David Gould
ISBN 1 85397 220 7 pbk 24pp £2.50

Rum and the Small Isles
Kathryn Goodenough & Tom Bradwell
ISBN 1 85397 370 2 pbk 48pp £5.95

Skye
David Stephenson & Jon Merritt
ISBN 1 85397 026 3 pbk 24pp £3.95

Scotland: the creation of its natural landscape
Alan McKirdy & Roger Crofts
ISBN 1 85397 004 2 pbk 64pp £7.50

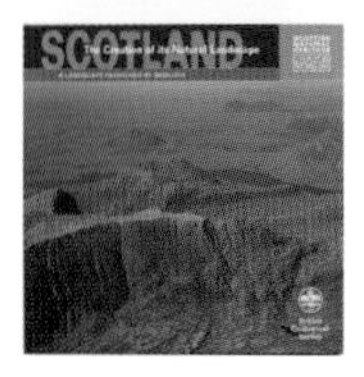

Series Editor: Alan McKirdy (SNH)
Other books soon to be produced in the series include: Ben Nevis and Glencoe, Western Isles

SNH Publication Order Form

Title	Price	Quantity
Arran & the Clyde Islands	£3.00	
Cairngorms	£4.95	
East Lothian & the Borders	£3.00	
Edinburgh & West Lothian	£4.95	
Fife & Tayside	£4.95	
Glasgow and Ayrshire	£4.95	
Glen Roy	£4.95	
Loch Lomond to Stirling	£2.00	
Mull and Iona	£4.95	
Northwest Highlands	£6.95	
Orkney & Shetland	£2.50	
Rum and the Small Isles	£5.95	
Skye	£3.95	
Scotland: the Creation of its natural landscape	£7.50	
Postage and packaging: free of charge within the UK. A standard charge of £2.95 will be applied to all orders from the EU. Elsewhere a standard charge of £5.50 will apply.		
	TOTAL	

Please complete in **BLOCK CAPITALS**

Name ______________________________

Address ______________________________

____

____

Post Code ______________________________

Method ☐ Mastercard ☐ Visa ☐ Switch ☐ Solo ☐ Cheque

Name of card holder ______________________________

Card Number ☐☐☐☐ ☐☐☐☐ ☐☐☐☐ ☐☐☐☐

Valid from ☐☐ ☐☐

Expiry Date ☐☐ ☐☐

Issue no. ☐☐

Security Code ☐☐☐ (last 3 digits on reverse of card)

Send order and cheque made payable to Scottish Natural Heritage to:
Scottish Natural Heritage, Design and Publications, Battleby, Redgorton, Perth PH1 3EW
pubs@snh.gov.uk

www.snh.org.uk